Galileo was Not Really Right

BY

ELIPHAS PHIRI

MttC Publications

DEDICATION

To my daughter Zoe

Table of Contents

PREFACE

For 400 years we have believed that objects fall at the same rate due to Galileo's famous experiment at the Leaning Tower of Pisa. But as I was revising my theory on the cause of gravity, one simple formula and Newton's Law showed that Galileo was not right; objects do not fall at the same rate. subsequently, I also realised that the experiment was not accurate enough and not adequate enough.

EP 01/03/23

PART 1

CHAPTER 1

REDEFINING ELECTROSTATICS

Present Electrostatic Theory

High school science tells us that there exist two charges; the positive and the negative. The like charges repel and the unlike charges attract.

These are defined arbitrary according to the object rubbed and by the type of cloth used to rub it by.

The glass rod rubbed with silk is said to be positively charged. The rubber rod rubbed with wool is negatively charged.

The theory to explain this phenomenon is atomic and electron based. It says that the transfer of electrons is the cause of the electrostatic phenomenon. A neutral object has equal numbers of both positive and negative charges so that its overall charge is zero. When rubbed, electrons are rubbed off the object onto the other. So one either has excess or deficit of the electrons. Since the electrons are negatively charged, the object with excess electrons gains a negative charge while the one with a deficit of electrons gains a positive charge, since the protons are positively charged.

Fields

The field for a positive charge was arbitrarily taken to be radial away from the charge or charge carrier, and that of the negative charge radial but toward the charge. This concept is used to explain attraction and repulsion.

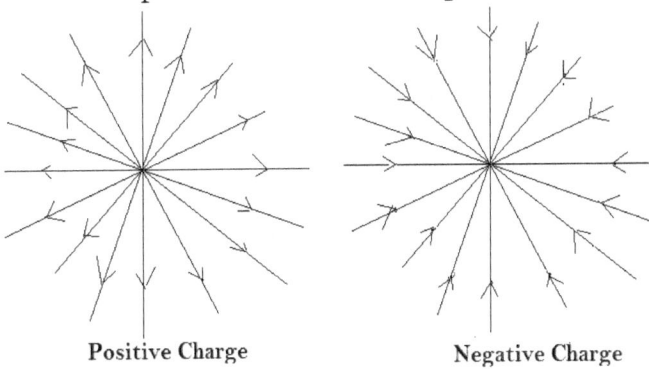

Positive Charge   **Negative Charge**

Definition

A charge is defined as that which gives rise to electricity. Since it gives rise to electricity when it moves, a charge must be considered as a piece of electricity or energy.

Charge must be considered as a qualitative quantity like mass, and not as a particle.

The overall field seen is of the charge carrier not the charges themselves. In short, we don't know the shape of the field. The field is shown as an overall shape of the charge carrier. I therefore propose a 'real' shape of the field of the electron.

Oersted's Experiments

Oersted noticed that a current carrying wire develops a magnetic field around it. This field can be physically observed by sprinkling iron fillings around the wire.

The field was found to be circular all around the wire with the wire as the center.

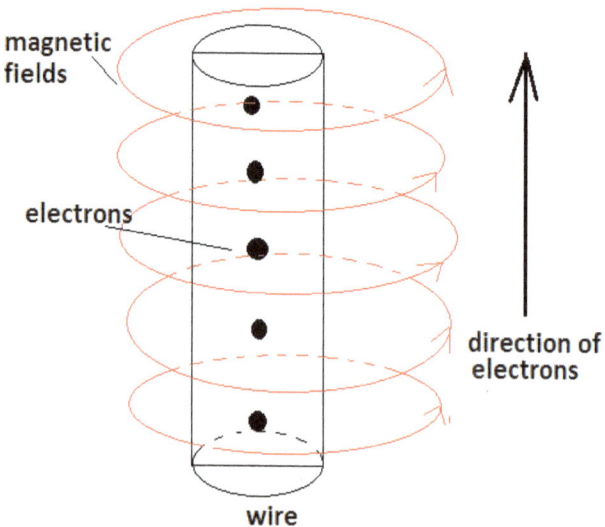

Redefining Charges and Electric Fields

Oersted's phenomenon happens only when the current is flowing. This means that there is a phenomenon which makes the electrons flow and hence create a magnetic field. But the electrons possess a field, a negative field, whether it is moving or not. I propose that the shape of that field around the electron is circular and anticlockwise when the current is moving north as per right hand rule.

This is like taking a slice of the cylindrical electric field around a current carrying wire. When the current is off, the slices of fields point in random directions as per domain theory of magnetism. The potential difference created by a wire connected to a battery makes these slices of fields to align in one direction and fall/move into that field, analogous to an object falling into a gravitational field. The potential difference provides a bigger field which attracts other fields and align them according to the center of mass.

The center of mass is pulled towards the earth in a gravitational field so that a falling object faces the earth or has a face. In the same way an electron has a face. Its center of mass points toward the field. The face is the north pole of the wire or solenoid.

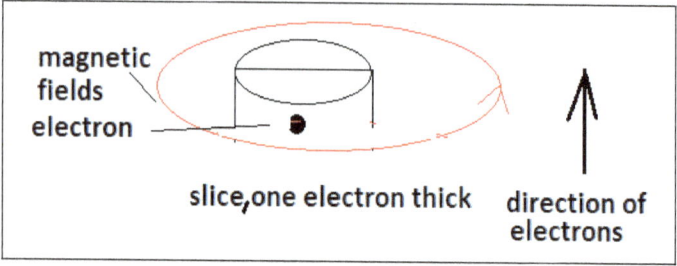

An electron has been proved to be a piece of mass so it has a center of gravity/mass. This is the centre of mass which twists in an electric field to face north or in a gravitational field to face down.

When current is off, the center of mass twists to other directions determined by the electrostatic forces of the other particles around it and gravity. Hence the electrons point in random directions. The nucleus scrambles the electrons.

The shape of the fields around a charge or charge carrier has been given as radial for both negative and positive charge.

The 'real' shape of the electric field has not been conclusively observed. Even in the case were it is photographed as radial, it can be argued that it shows the path of the charged particles not the field itself, just as a piece of nail will be radially attracted to the magnet though the field of the magnet is actually oval and not radial.

Slicing through the magnetic field of a current carrying wire at electron level, we can assume that the field we get is the 'real' field of the electron.

The field for positive or proton can then be said to be clockwise when the proton is moving north. Positive and negative of a charge is now considered wholly in a directional sense and not arbitrary and in terms of which object is rubbed by a particular object.

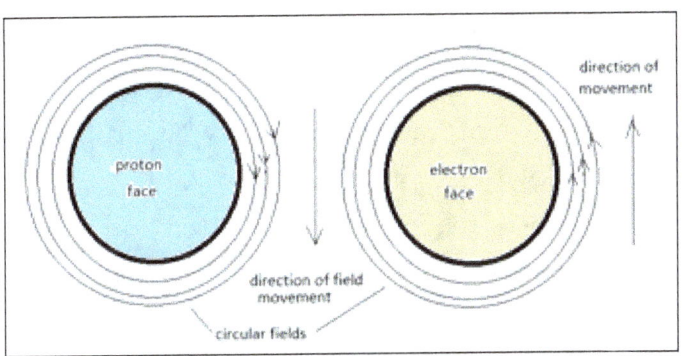

The positive charge carrier is the proton i.e. a piece of mass with a field in clockwise direction when it is facing or moving south. A negative charge carrier is a piece of mass with a field moving anticlockwise when it is moving in the north direction i.e. in opposite direction.

There is only one charge, therefore, which manifests as either positive or negative according to the direction in which the mass is spinning and moving. The definition is based on the direction of the spinning in relation to the direction of the forward movement.

Instead of calling charges positive and negative, we can talk of positive and negative charge carriers. The electron is the negative charge carrier and the proton is the positive charge carrier . But they are only negative or positive when moving north while rotating anticlockwise and vice-versa.

Now the mechanics of attraction and repulsion will change. Attraction and repulsion will be according to how the particles are oriented next to each other. It will be like in two magnets next to each other. In short, the electron and proton effectively become flat circular magnets [pancake magnets].

Two Current Carrying Wires Parallel Next to Each Other

Two current carrying wires with current in the same direction attract each other as the diagram and fields show. The fields merge so the two wires attract.

Two wires with current in the opposite direction repel each other as the diagrams show. The fields collide so the two wires repel.

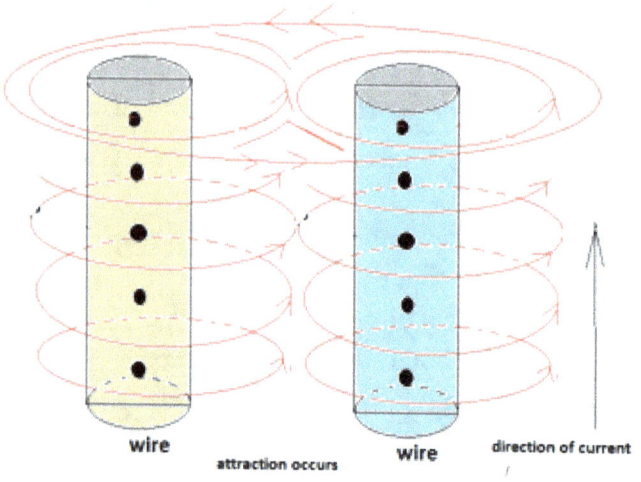

wire   attraction occurs   wire   direction of current

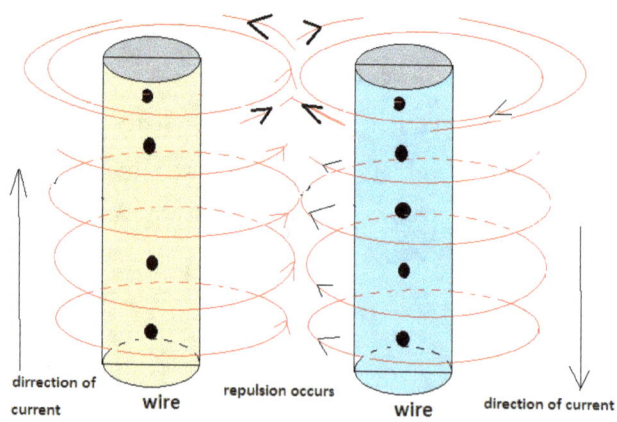

dirrection of current   wire   repulsion occurs   wire   direction of current

But oriented face to face i.e. north to north we would expect the wires or solenoids to repel each other. That is; it will be like a solenoid with north-pole facing another

10

north-pole.

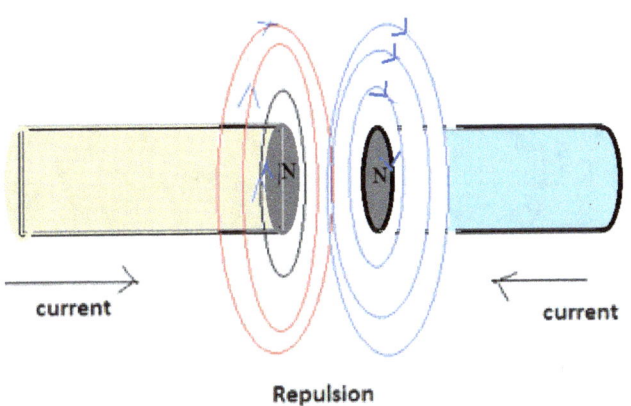

Repulsion

Sideways Attraction and Repulsion

So now an electron and proton can be either attractive or repulsive next to another electron or proton depending on how they face each other. Electrons in the same direction next to each other will attract sideways. The same goes for protons. Electrons next to each other in opposite directions will repel. The same goes for proton.

Electrons and protons next to each other in the same direction will repel. Electron and proton next to each other in opposite direction will attract.

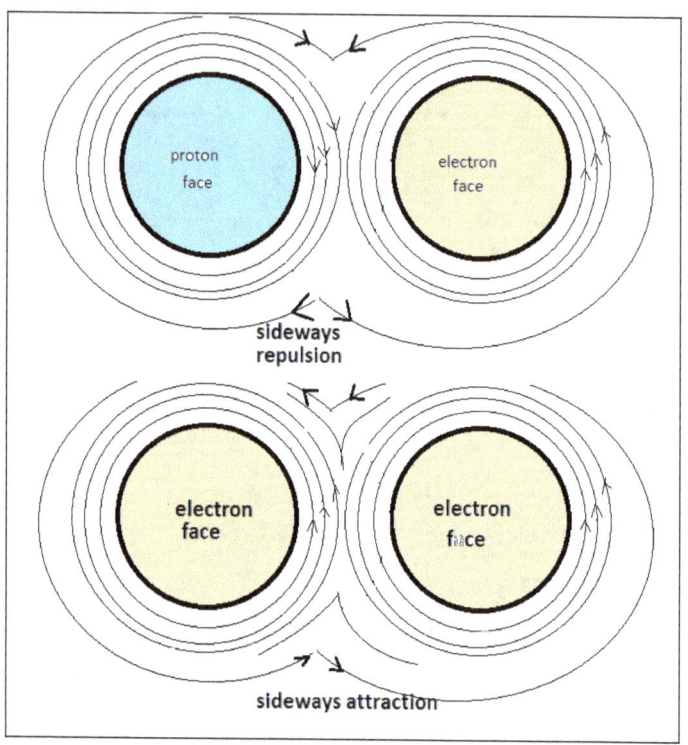

Face to Face Repulsion

If they face each other face to face, they will repel, and if they face each other face to back, they will attract. The electron will also behave the same.

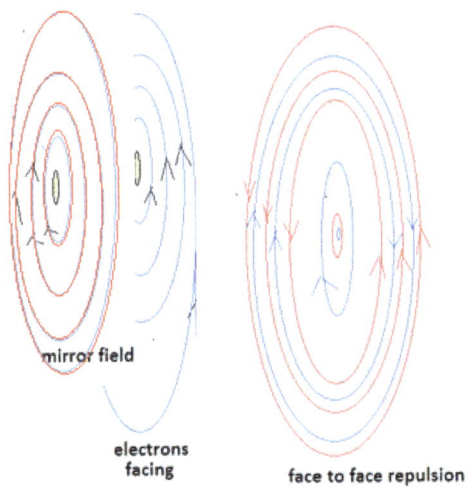

mirror field

electrons
facing

face to face repulsion

When the electrons face each other face to face, the fields will face in opposite directions. One will be a mirror image of the other.

The two fields therefore collide and repulsion occurs.

Face to Back Attraction

When the electrons are oriented face to back the fields will be facing each other all in anticlockwise direction i.e. in the same direction.

The fields therefore merge as in magnets of different poles facing each other.

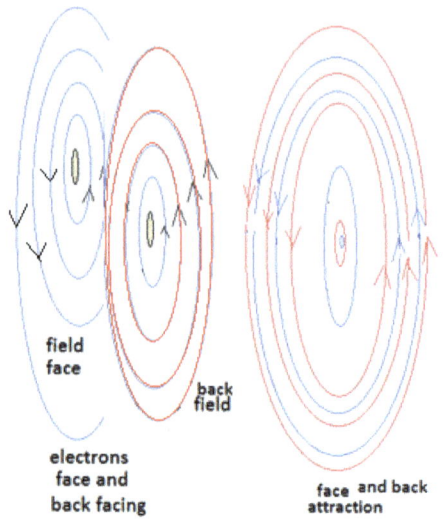

field
face

back
field

electrons
face and
back facing

face and back
attraction

Coulomb's Law

Coulombs law still applies to all the cases because essentially, the field is between electrons and protons and so is electric.

Proof from Chemical Reactions

The chemistry of reactions says that bonds are a result of electron sharing. But if the electron is wholly repulsive to the other electron then this may not occur or may require tremendous force since as the distance between the electrons reduces the Coulomb force becomes great.

Chemical reactions must occur, therefore, because the electrons orient themselves in such a way that attraction occurs. After the initial attraction of the differently charged atoms, the electron sharing must be achieved by the attraction of the valence electrons.

 The electrons are on the surface and the protons are inside. As the atoms come closer to each other, the electrons get closer. So the electrons will reach a point where their repulsive force is greater than or equal to net pull of the atom. So electron sharing or pairing can only occur by electron-to-electron attraction.

This will require less energy than in a repulsive situation.

Then once the valence pairs are set up, the pairs will be oriented in such a way that they repel the next pair resulting in the VSEPR model of molecular structure.

For instance, the compound Beryllium Chloride [$BeCl_2$] can be said to form a bonding pair by the face-to-back attraction of the valence electrons. Then these pairs oriented themselves in such a way that repulsion occurs i.e. one face of the pair now faces another face of the other pairs or the back of one pair now faces the back of the other pair or the pairs being side to side but in opposite directions.

Electron pair set up by two electrons atrracting

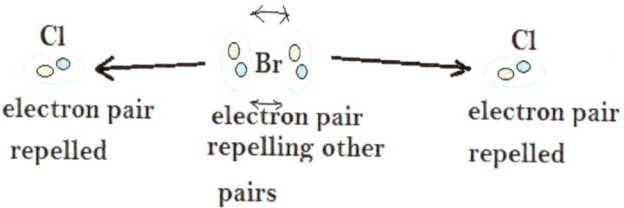

electron pair repelled

electron pair repelling other pairs

electron pair repelled

CHAPTER 2

THE CAUSE OF GRAVITY

Polarized Surfaces

The experiment that shows two current carrying wires in the same direction/opposite directions attracting/repelling each other can be applied to the elusive understanding of the cause of gravity.

When the current is on, the wires effectively become magnets and either attract or repel. When the current is off, the wires do not, attract or repel. Nevertheless, the wires can be taken as infinitely many electrons oriented in random directions in the neutral wire i.e. when the current is switched off.

Though the object is neutral, it is slightly negative due to the difference in the positions of the electrons and the protons as per atom theory. The electrons are at the surface and the protons are at the center. In short, all neutral matter is slightly negative or polarized as in polarization of dielectrics in capacitors. The electrons point in random directions so that their fields do not add-up to give a greater field and hence force.

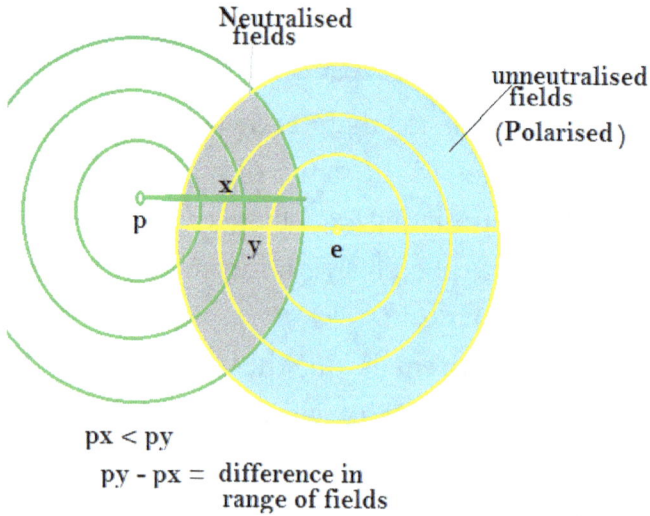

Neutralised fields

unneutralised fields

(Polarised)

x

p

y

e

px < py

py - px = difference in range of fields

If x is the radius that positive proton field covers, then the fields are neutralized at the regions were it intersects the fields of the electron up to the end of x, then from x to end of y and beyond the fields are not neutralized i.e. all the space 2y ⁻ x. This region is polarised. It is slightly negative.

Also since the field strength reduces with the factor of $1/r^2$, the field strength at a point P will less for the proton than for the electron because of their difference in position. The electron is nearer to P than the electron though they are in the same atom.

So a neutral surface will be slightly negative. Some of the fields will be pointing up, giving rise to the gravitational force and some of the fields will be lying horizontal, giving rise to the electrostatic force.

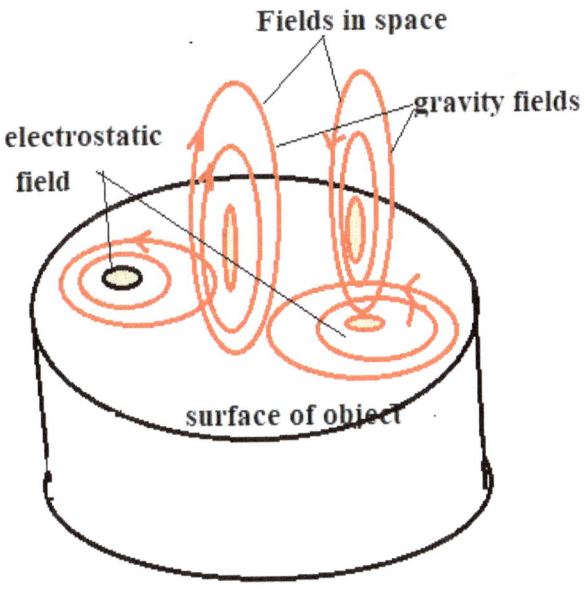

So the two or more neutral objects facing each other are like two wires with current in the same direction. The fields in specific directions will, therefore, hook up just as the two wires attract. The two objects will attract. This is the force we call gravity. It is simply another manifestation of the electric field just as magnetism is a manifestation of the electric field. Other fields will be repelling though. Even if they are vertical and hence gravitational, they will repel if oriented in repulsion position.

But the electrostatic force will also be happening at the same time. The horizontal fields will be attracting or repelling the other horizontal fields on the opposite surface too.

For the fields that are vertical the gravitational force will act upward and the electrostatic force will act sideways. For the fields that are horizontal, the electrostatic force will act upward and the gravitational force will act horizontal. This solves the dilemma of the of how a gravitational field could be induced by an electron when the principle is that electron repels another electron. By redefining positive and negative charge in terms of direction of the field, we can surmount the elusive cause of gravity.

The electric field is therefore the true grand unifying field. The electron is the true boson. All particle physics can now be reduced to just two particles; the electron and the proton. There is now no need to invent a particle that imparts gravity or mass. That makes such a particle a fable. The boson is a fable and unnecessary.

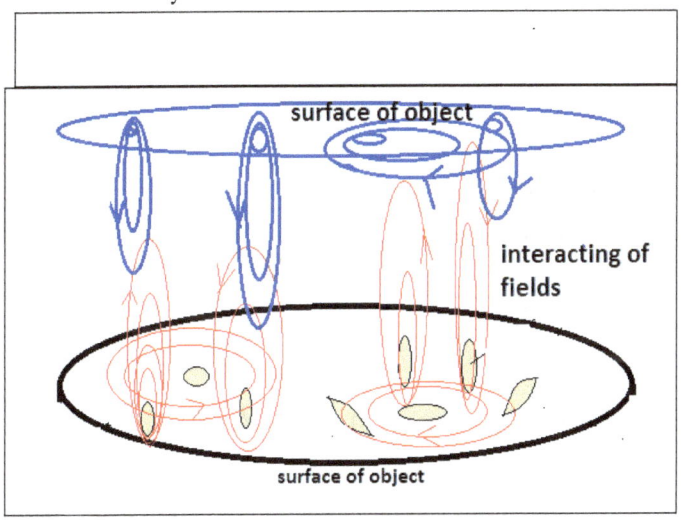

surface of object

interacting of fields

surface of object

Factor of Polarisation, Я

The surfaces of all neutral objects are slightly charged and the charge is negative. The electron and the proton are in different positions in the atom. This is what polarizes neutral objects. The positive clockwise field of the proton do not align with the negative anticlockwise field of the electron. The electron field overlaps slightly so that there is no total cancellation of the fields. The neutral matter is therefore left slightly negatively charged.

We can introduce a factor to show the amount of Polarization. That factor can be the Coptic symbol Я which can be read as 'backward R'.

The polarising factor, will apply to all masses. It shows the amount of polarisation quantitatively.

Equating the Gravitational Force to Electric Force

The gravitational field at the surface of the earth is given as m = $\dfrac{F_g}{g}$, from F = ma. The electric field due to a source charge, q , is given as $E = \dfrac{F_c}{q}$.

For neutral but polarised matter the electric field can now be given as ;

$$\boxed{E = \left|\dfrac{F_c}{Яq}\right.}$$

This field is equivalent to the gravitational field at the same point. It can be called the Gravito-Electrical Field and denoted by E_g.

$$\text{so } E_g = \left|\dfrac{F_c}{Яq}\right.$$

Equating the gravitational field to the gravito-electric field we get;

$$m = E, \quad \dfrac{F_g}{g} = \dfrac{F_c}{Яq} \quad , \quad \left|F_g\right| = \dfrac{gF_c}{Яq}$$

But Why is Gravity Weak and Electric Force Strong

This force, gravity, is weak because most of it is canceled out by the protons. The difference in the distance between the protons and electrons makes the surface slightly polarized or negative. Because this polarization is small, the gravity is also small.

The electrons on the surface face in random directions and we would expect some of them to cancel each other out and some to repel. This weakens the force further. However, the protons hold enough electrons in molecules in the same direction more than those that are in random direction. There is therefore a net attractive force between the objects facing each other. There will be enough pairing of them to attract though they maybe in random directions.

Another reason why the electrostatic force maybe greater than gravity is that when electrons face each other face to face or face to back there is more of the field interacting than when they face sideways. There is more surface area when they meet surface to face than sideways. Only the part of the sideways fields where they meet will interact, and that part is small almost a point.

The static force is greater than gravity because

the brushed off electrons orient themselves in one direction with either their faces or bottoms facing up

So the net force is slightly greater than zero. Though some will miss each other, most of them will hook up like in a Velcro zipper.

Redefining Electrostatic Charge of Objects

When rubbed, electrons on the surface of the neutral object are loosed from electrostatic pull of the protons, analogous to wearing off of particles due to friction. These electrons now face up like a magnet's North Pole or South Pole facing up from the surface of the neutral object.

The electrons in the other object have turned upside down so that their bottoms are facing up.

The objects with electrons facing up can be said to be negatively charged. The objects with electrons facing down can be said to be positively charged. Both negativity and positivity being caused by the electron.

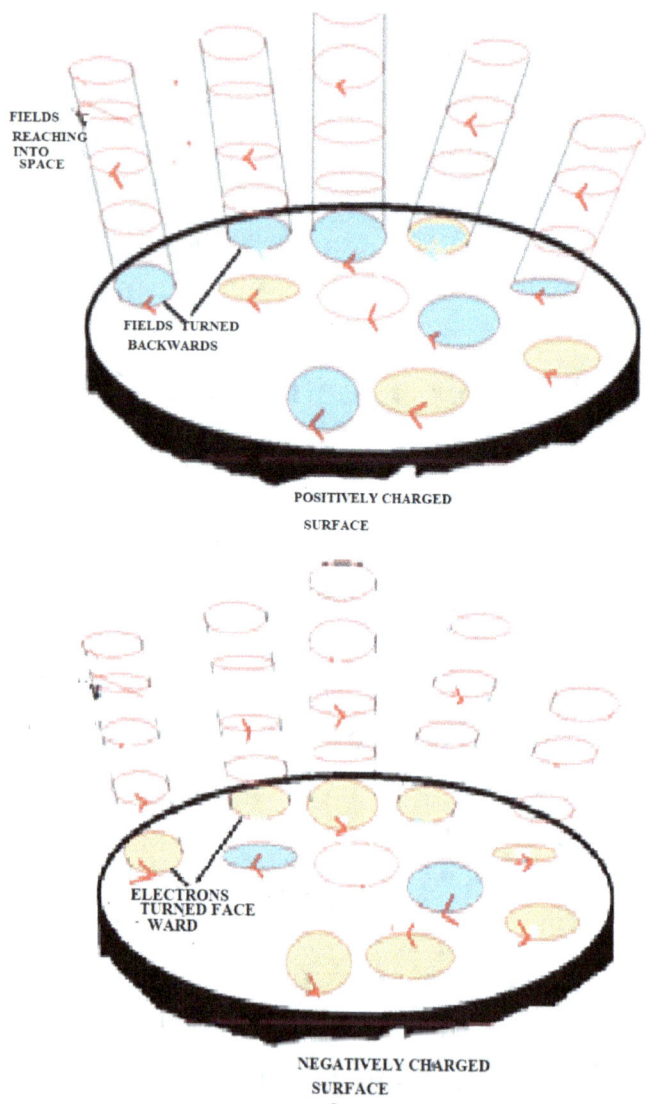

The loosed electrons, under gravity, settle in a

stable position either facing up or down or sideways, like a rectangle which can settle in a stable equilibrium either vertical or horizontal under the influence of gravity.

The field on the face of the electron or proton is like that of a North Pole or South Pole of a magnet. Two objects of like charge facing each other is like having two poles of a magnet facing each other. The electrons are facing each other face-ward, so they repel as discussed in chapter 1.

Two differently charged objects will be like two different magnetic poles facing each other.

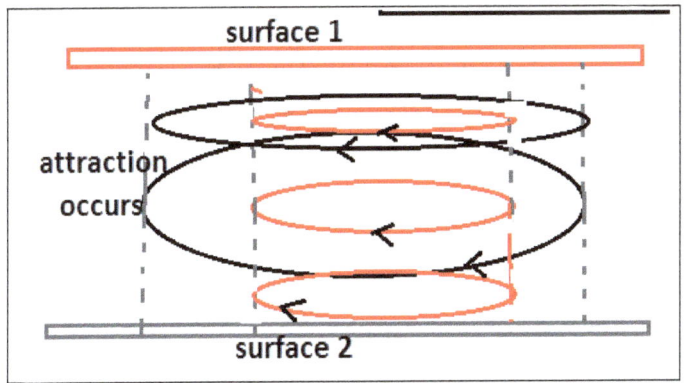

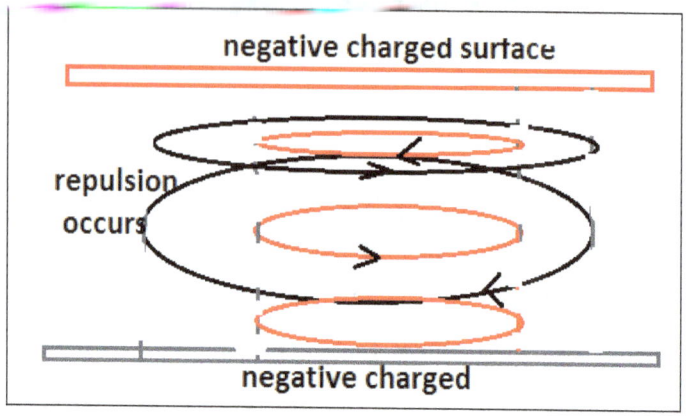

This re-interpretation and theory of electrostatics does away with the jumping of electrons from one object to another, which is more like a chemical reaction, in explaining electrostatic phenomena.

This also helps in explaining positive polarization intuitively. The protons are deep inside the atom so that explaining positive charge on a surface with the idea of electrons being pushed away is problematic. Were exactly will the electrons move inside the atom so that it becomes positive?

Earthing

These loosed electrons are weakly held by the rest of the body so they can flow to the earth or from the earth.

Interaction of Electric and Gravity Field

The charged objects exert gravity and experience gravity because some of the electrons under the loosed electrons are oriented with faces away from the surface so that the rim to rim orientation fields is still happening.

The gravitational field and hence the gravitational force and the electric field and hence its force interact or interfere with each other just like any other force. They can interact repulsively or attractively, and then the net force will be;

$$F_n = \sqrt{F_g^2 + F_e^2}$$

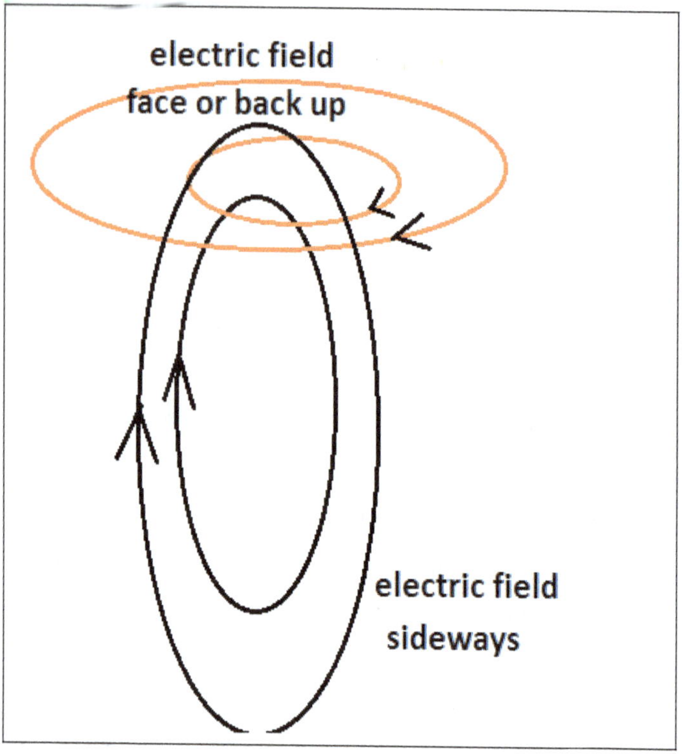

electric field
face or back up

electric field
sideways

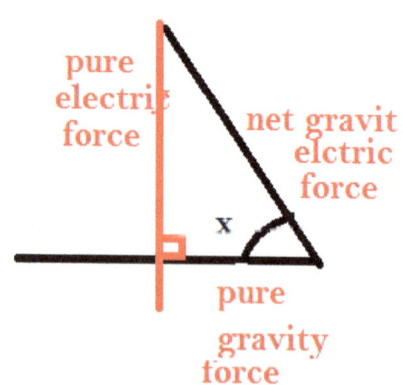

pure
electric
force

net gravit
elctric
force

x

pure
gravity
force

Net gravito-electric force horizontally = gravity force = $F_g{}^2 \cos x$

Net gravito-electric force vertically = electric-force = $F_e{}^2 \sin x$

i.e.
$$F_{ge} = \sqrt{F_g{}^2 \cos x + F_e{}^2 \sin x}$$

i.e.
$$F_{ge} = \sqrt{[G\frac{M_1 M_2}{r^2}]^2 \cos x + [K\frac{Q_1 Q_2}{r^2}]^2 \sin x}$$

But these fields are usually oriented at right angles to each other; its either the field is vertical or it is horizontal. The Gravito-electrico force, therefore, reduces to gravity force or electrostatic force depending on where the field is pointing i.e. vertical or horizontal on the ground.

$$F_{ge} = \sqrt{[G\frac{M_1 M_2}{r^2}]^2 \cos 90° + [K\frac{Q_1 Q_2}{r^2}]^2 \sin 90°}$$

$$F_{ge} = \sqrt{[G\frac{M_1 M_2}{r^2}]^2 [0] + [K\frac{Q_1 Q_2}{r^2}]^2 [1]}$$

$$F_{ge} = \sqrt{[K\frac{Q_1 Q_2}{r^2}]^2} \quad , \quad F_{ge} = K\frac{Q_1 Q_2}{r^2}$$ When the field is horizontal.

OR

$$F_{ge} = \sqrt{[G\frac{M_1 M_2}{r^2}]^2 cos90° + [K\frac{Q_1 Q_2}{r^2}]^2 sin90°}$$

$$F_{ge} = \sqrt{[G\frac{M_1 M_2}{r^2}]^2[1] + [K\frac{Q_1 Q_2}{r^2}]^2[o]}$$

$$F_{ge} = \sqrt{[G\frac{M_1 M_2}{r^2}]^2[1]} \quad , \quad F_{ge} = G\frac{M_1 M_2}{r^2}$$ when the field is vertical.

Equating Coulomb Constant to Gravitational Constant
i) For Electron to Electron

The electrostatic force of the electrons standing side to side is also the gravitational force since we said it's the source of the gravity. So for electrons facing each other face to face the force is repulsive and negative i.e.

$$F_C = -K\frac{Q_1 Q_2}{r^2} .$$

Since it's repulsive, this force cannot be

considered as gravitational but electrostatic. Also for vertical fields whose directions collide and hence the force becomes repulsive, the force is not gravitational even if the fields are vertical. Or it can be called repulsive gravitation or negative gravitation.

The attractive force will be considered as the gravitational force and must act sideways. Since it is considered gravitational the polarizing factor for electrons $Я_{ee}$ must be introduced.

So the Coulomb force for fields oriented vertical to vertical will be; $F_{ee} = K \dfrac{Я_{ee}Q_1Q_2}{r^2}$

instead of $K \dfrac{Q_1Q_2}{r^2}$. This is the gravitational force.

This polarizing factor in this case can be considered as the result of the difference in distance between when the electrons are face to back and when they are side to side. The greatest attraction occurs face to back because more fields merge than sideways. There is more surface to interact face-to-face than sideways. There are fewer fields sideways than face-ward i.e. the thickness of the field is far much smaller than the width of the field. It is a pancake field or donut field. A slice of the fields one electron thick will merge sideways.

The face-face interaction will make a cylinder implying more fields per area and the side to side will make a plane implying less field per area, for the same distance from the center of the electron.

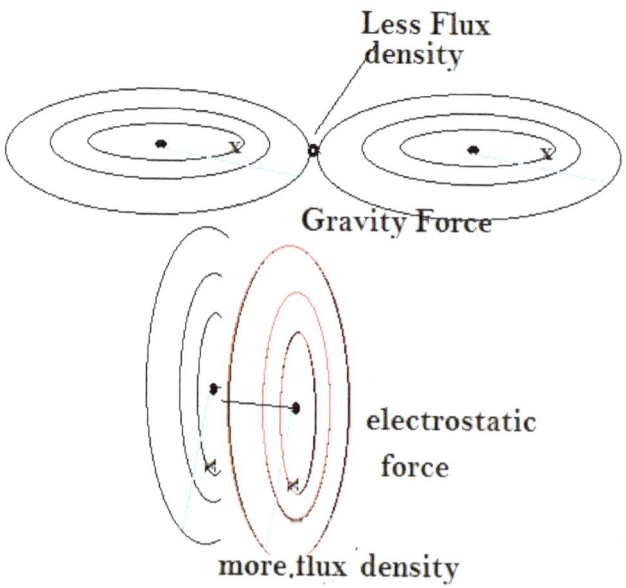

So Яee is determined by the difference in the

flux density of the fields when the attraction is sideways and when the attraction is face-to-face i.e. $Я_{ee} = \phi_{E2} - \phi_{E1}$, $\phi_{E2} > \phi_{E1}$. For proton to proton, it will be $Я_{pp} = \phi_{E2} - \phi_{E1}$. For electron to proton it will be $Я_{ep} = \phi_{E2} - \phi_{E1}$.

Equating $K \dfrac{Я_{ee}Q_1Q_2}{r^2}$ to $G \dfrac{M_1M_2}{r^2}$, the Gravitational force, we get;

$$K \frac{Я_{ee}Q_1Q_2}{r^2} = G \frac{M_1M_2}{r^2}$$

$$G = K \frac{Я_{ee}Q_1Q_2}{M_eM_e} , G = K \frac{Я_{ee}Q^2}{M^2}$$

$$G = K \, Я_{ee} z^2$$

$$\boxed{G = Я_{ee} K z^2}$$

But Q/M (z) is charge to mass ratio and was found by J.J. Thomson to be $1.76 \times 10^5 C/kg$.

Magnitude of Я for Particle to Particle

i) for electron to electron attraction;

$K\pi^2 = (1.76 \times 10^8)^2$ C/kg(8.99×10^9) Nm^2/C^2

$= 27.8 \times 10^{25}$

So G $=$ Я(27.8×10^{25}), but G$=$ 6.67×10^{-11}Nm^2/kg^2

So $^Я ee = 6.67 \times 10^{-11}/27.8 \times 10^{25}$,

$^Я ee = 0.24 \times 10^{-36} = 2.4 \times 10^{-37}$

For electron to proton $^Я{}_{ep} =$ 4.40×10^{-40}N^{2}m^4/Ckg^4. This is the $^Я{}_{ep}$ of the hydrogen atom.

This occurs when the two electrons are oriented sideways and pointing in the same direction as the diagrams show.

When the electrons are facing, face to face, [north-north] or back to back, then repulsion will occur. If they are facing face to back attraction will occur. This attraction or repulsion is wholly electrostatic since the fields will fit into each other face to face or face to back leaving no distance between them or their centers.

So now the electron to electron attraction is not wholly repulsive as before, and proton to proton interaction is also not wholly repulsive as before. Both attraction and repulsion will occur according to the orientation of the particles involved.

ii) Electron to Proton

The greatest attraction occurs face to face with centres of mass touching. A proton to electron Polarising Factor, $Я_{EP}$, must therefore be introduced just as in the first case of the electron to electron attraction.

This polarizing factor in this case can be considered as the result of the difference in distance between when the electrons and proton attraction is face to face and when they are side to side.

So the Coulomb force will be;

$$F_e = -K \frac{Я_{ep}Q_1Q_2}{r^2} \text{, instead of } K = -\frac{Q_1Q_2}{r^2} \text{,}$$

where r is the distance of the protons from

the electrons i.e. radius of the atom.

For a Point P

For an object O at point P the field at that point from the electron will be $E_{eo} = k Я_{eo} \dfrac{-Q}{x^2}$, and from the proton, $E_{po} = k Я_{po} \dfrac{Q}{(r+x)^2}$, r = radius of the atom, and x distance from proton to P.

The difference in the fields of the protons and electrons at the surface of the object, i.e. the polarization, will determine the resultant field and hence force experienced by a particle at a point P.

Since the electrons are generally at the surface, the radius of the atom can be taken as the difference in length between the electrons and protons.

So the net field due to difference in length in that direction at that point will be;

$$E_n = E_p + E_e$$

i.e. $E_n = kЯ_{po}\dfrac{Q}{(r+x)^2} + kЯ_{eo}\dfrac{-Q}{x^2}$

$E_n = kQ(\dfrac{Я_{po}}{(r+x)^2} - \dfrac{Я_{eo}}{x^2})$, this electric field in this orientation is now called the Gravity Field, E_g.

So $\boxed{E_g = kQ(\dfrac{Я_{po}}{(r+x)^2} - \dfrac{Я_{eo}}{x^2})}$

So the electric force experienced by a test charge q /mass O at P in that direction is ;

$F_e = E_g q = kQq(\dfrac{Я_{po}}{(r+x)^2} - \dfrac{Я_{eo}}{x^2})$, but this is the pure gravitational force because of the orientation.

So $\boxed{F_g = E_g q = kQq(\dfrac{Я_{po}}{(r+x)^2} - \dfrac{Я_{eo}}{x^2})}$

The Gravitational Force at P Near the Ground

This electric field on the test mass/charge at

this point and near the ground from a single electron is equal to the gravitational field at this point.

That is; $\quad kQq(\dfrac{Я_{po}}{(r+x)^2} - \dfrac{Я_{eo}}{x^2}) = \dfrac{F_g}{g}$

So $F_g = kQqg\left(\dfrac{Я_{po}}{(r+x)^2} - \dfrac{Я_{eo}}{x^2}\right)$, This is the Gravity Force experienced by a charge/object O near the ground because of g.

If many electrons are acting on P then

$F_g = kNQqg\left(\dfrac{Я_{po}}{(r+x)^2} - \dfrac{Я_{eo}}{x^2}\right)$, N being the number of electrons attracting O at P.

For two objects near each other with gravitational fields g_1, g_2 then the net Force will be; $F_g = 2kNQq\left(\dfrac{Я_{po}}{(r+x)^2} - \dfrac{Я_{eo}}{x^2}\right)(g_e + g_o)$

Or $F_g = 2kNQq\left(\dfrac{Я_{po}}{(r+x)^2} - \dfrac{Я_{eo}}{x^2}\right)g_n$, $\quad g_n = g_e + g_o$, i.e. g_n = net g = g.

But two masses or charges will also be interacting pure-electrostatically repulsively or attractively. This electrostatic force is given by

the classic formula; $F_e = \dfrac{K\frac{Qq}{r^2}}{}$ if attractive

or $\qquad\qquad F_e = \dfrac{-K\frac{Qq}{r^2}}{}$ if repulsive.

So the net gravito-electrico force, F_{ge}, near the surface of the earth is;

$$F_{ge} = \sqrt{F_g^2 cosx + F_e^2 sinx}$$

$$F_{ge} = \sqrt{[2kNQqg\left(\frac{Я_{po}}{(r+x)^2} - \frac{Я_{eo}}{x^2}\right)]^2 cosx + [K\frac{NQq}{x^2}]^2 sinx}$$

This is the force an object near a planet or earth experiences. It is not purely gravity, but the net of gravity and electrostatic forces.

iii) Polarising Factor for Neutral Objects (Equating G to K)

For two neutral objects of masses M_1 and M_2 separated by a distance of x meters. The two objects can be considered to be charged objects with charges Q and q. The Gravito-Coulomb force can therefore apply to the objects.

The force of gravity between them is given as;

$$F_g = G\,\frac{M_1 M_2}{x^2}$$

Equating the two we get; $F_c = F_g$

$$kQq(\frac{Я_{po}}{(r+x)^2} - \frac{Я_{eo}}{x^2}) = G\,\frac{M_1 M_2}{x^2}$$

$$G = \frac{KQq}{M_1 M_2}(\frac{Я_{po}}{(r+x)^2} - \frac{Я_{eo}}{x^2})x^2$$

$$G = \frac{KQq}{M_1 M_2}\left(\frac{Я_{po}x^2}{(r+x)^2} - Я_{eo}\right)$$

For the whole object the total polarised charge will be number of electrons oriented for attraction multiplied by the charge.

$$\text{So}\quad G = \frac{KNQq}{M_1 M_2}\left(\frac{Я_{po}x^2}{(r+x)^2} - Я_{eo}\right)$$

N is number of pairs of electrons oriented for attraction.

Not all electrons will be oriented in such a way that attraction occurs. Some will repel each other and others will be oriented perpendicular to the sideways field so that not attractive nor repulsive force will occur. So it is possible to have more than half of the surface electrons not oriented in attractive way but still have a net attractive force.

But Q/M ,$\not{Z}$, [z cross] can be called the polarised charge to mass electron density or charge to mass ratio.

$$\text{So} \quad G = KN\not{z}_1\not{z}_2\left(\frac{Я_{po}x^2}{(r+x)^2} - Я_{eo}\right)$$

$\not{Z}$ is equal to the charge to mass ratio. But because N is a fraction of the total number of charges, the overall charge is smaller compared to the charge to mass ratio of the electron to electron attraction. This will reduce the Coulomb force in comparison to the electron to electron force.

The value $N\left(\dfrac{Я_{po}x^2}{(r+x)^2} - Я_{eo}\right)$ can be called the Polarisation factor of neutral objects, Я.

So $\qquad Я = N\left(\dfrac{Я_{po}x^2}{(r+x)^2} - Я_{eo}\right)$ and

$$\boxed{G = KЯ_{z_1}}z_2$$

The Net Gravito-Electrico Field between Two Neutral Objects

For two neutral objects of masses M1 and M2 separated by a distance x with N pairs of electron to electron attraction, the net field will be

$E = E_2 - E_1,$

$E = kQN(\dfrac{Я_{po}}{(r+x)^2} - \dfrac{Я_{eo}}{x^2}) - [-kQN(\dfrac{Я_{po}}{(r+x)^2} - \dfrac{Я_{eo}}{x^2})]$

$E_g = 2\, kQN(\dfrac{Я_{po}}{(r+x)^2} - \dfrac{Я_{eo}}{x^2})$ and net force

$F_g = 2kNQq\left(\dfrac{Я_{po}}{(r+x)^2} - \dfrac{Я_{eo}}{x^2}\right) = 2kЯQq$

Or $\quad F_g = \dfrac{2G}{Яz_1z_2}NQq\left(\dfrac{Я_{po}}{(r+x)^2} - \dfrac{Я_{eo}}{x^2}\right) = \dfrac{2G}{z_1z_2}Qq$

Then there is the pure electrostatic forces acting on the masses too.

The net gravito-electrico force is,

$$F_{ge} = \sqrt{[2kNQq\left(\dfrac{Я_{po}}{(r+x)^2} - \dfrac{Я_{eo}}{x^2}\right)]^2 cosx + [K\dfrac{NQq}{x^2}]^2 sinx}$$

OR

$$F_{ge} = \sqrt{[2kЯQq]^2 cosx + [K\dfrac{NQq}{x^2}]^2 sinx}$$

OR

$$F_{ge} = \sqrt{[\dfrac{2G}{Яz_1z_2}NQq\left(\dfrac{Я_{po}}{(r+x)^2} - \dfrac{Я_{eo}}{x^2}\right)]^2 cosx + [K\dfrac{NQq}{x^2}]^2 sinx}$$

OR

$$F_{ge} = \sqrt{[\dfrac{2G}{z_1z_2}Qq]^2 cosx + [K\dfrac{NQq}{x^2}]^2 sinx}$$

This is the gravity experienced by masses far astronomically apart. This is the Universal Gravito-electrico Force.

Chapter 3
The Relativity of Mutually Attractive/Repulsive Objects

The Net Force of Mutually attracting Objects

For forces which are attracting, the forces add up. If the forces F1 and F2 are attracting each other the net Force F_n = F1 + F2 and not F1 - F2.

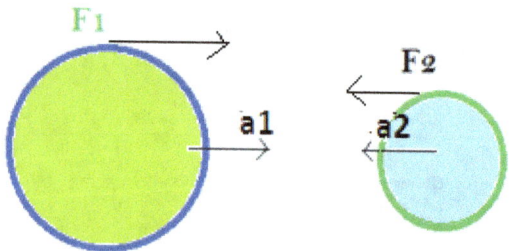

Suppose we have two objects attracting each other as shown, if the green object is kept still or fixed and the blue is left mobile. The force F2 will have a reaction force on the blue object in the same direction as F1 but equal to F2. The resultant force, therefore, will be the sum of the two forces i.e.

$$F_n = F1 + F2$$

The Net Force of Mutually Repulsive Objects

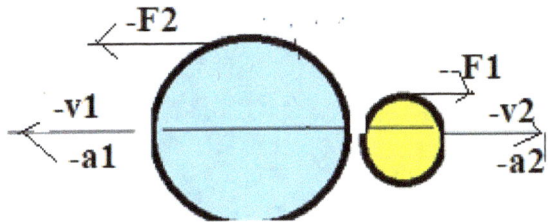

Similarly, for repulsive forces the Fnet will be the sum of the forces but negative in value. The direction of attractive forces can be taken as positive so that the repulsive force is negative. Net force, therefore, will be -F1 – F2,

$$F_n = - (F1 + F2)$$

Net Acceleration of Mutually Attracting Bodies

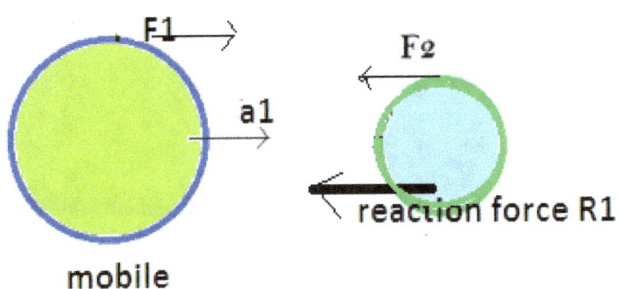

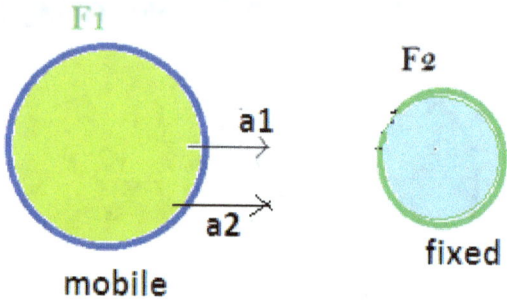

Suppose we have two objects attracting each other as shown, if the green object is kept still or fixed and the blue is left mobile, the effect of F1 will be to make the blue object move, and the effect of F2 will be to make the blue object move too. These movements will be in the same direction. The velocities and the accelerations will be sum of the two velocities and the two accelerations therefore at lower speeds less than the speed of light.

The net acceleration will therefore be; an = a1 + a2 i.e.

$$a_n = a_e + a_o$$

Net Acceleration of Mutually Repulsing Bodies

For repelling objects, the repulsive forces will

set up reaction forces R1 and R2 on the opposite objects. The net force will cause accelerations $-a_1$ and $-a_2$ on the two objects.

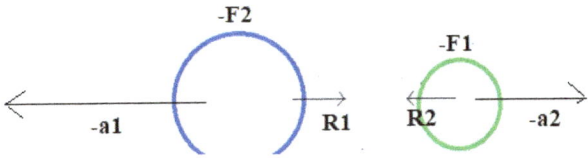

The net acceleration will therefore be the sum of the two accelerations i.e.

$$a_n = -(a_e + a_o)$$

This is the net acceleration that the object with the smallest force will move.

The Galilean-Relative Acceleration of Mutually Attracting Bodies

For two bodies in space each object is falling into another's gravity field at that object's g and the other object is also falling into that object with that object's g, they will meet faster and some way through the journey. The relative acceleration of one body from the observer on the opposite body, and vice versa, will be greater than if only one body has gravity and the other did not.

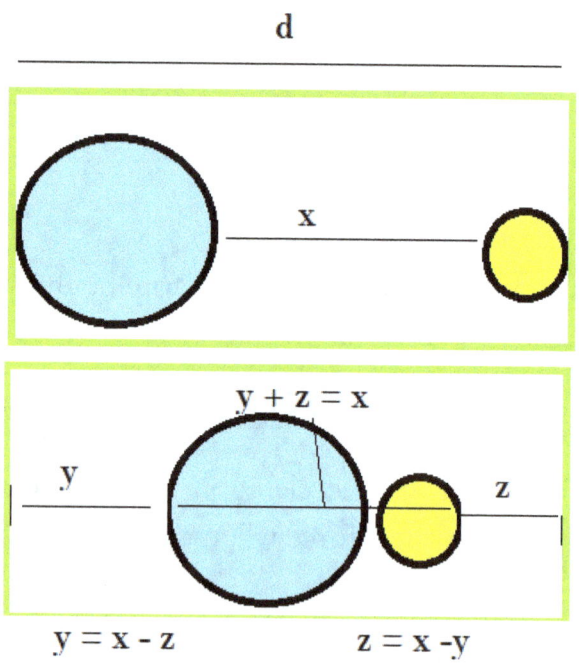

The blue object will fall into **z** by **y** meters while the yellow will move **z** meters. Instead of falling through the length **x**, the smaller object will seem to fall through **x − y** to an observer on the yellow object, and instead of falling through **y**, the bigger object will seem to fall through **x − z**.

Also instead of taking the time t_X to meet, the objects will take t_{X-Y} and t_{X-Z} to meet. So instead of the acceleration being $a_{xe} = V_1/t_x$,

it will be $a_{x-z} = \dfrac{V_1}{t_{x-z}}$. Since V is the same but $t_x > t_{x-z}$, a_{x-z} is greater than a_{xe}. Also the acceleration of object a_{xo} will be different from a_{x-y}.

It is like jumping up to catch a falling ball or object; the object is moving towards you at about 9.81 m/s² and you are also accelerating towards the object. You therefore meet the object earlier than if one did not jump. The net g of the jumping person and the falling object is therefore greater than 9.81m/s² for an observer in space who does not see the ground as a reference point of the two motions. To that observer objects fall at different rates.

So for an observer on the yellow object, oblivious to his movement towards the blue object, will see the blue object fall faster than normal and vice-versa.

If the two objects are repelling then the relative velocity noticed by an observer on one object at lower speeds will be the sum of the two velocities as they recede from each other i.e. $\mathbf{a} = \dfrac{-(v1 + v2)}{t}$ instead of $\mathbf{a} = = \dfrac{-v1}{t}$. Since the time is the same in both cases but the observer notices one velocity as sum of the two velocities the net acceleration is bigger than expected.

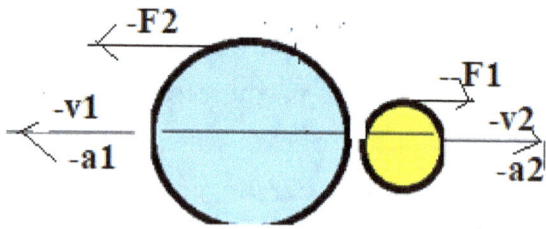

The Special-Relativity Acceleration Addition Formulae

For speeds near the speed of light the velocity addition of the repulsive objects reduce to that of the Lorentz Velocity Transformations equations; i.e.

$$U_x^{'} = \dfrac{u - v}{1 - \left(\dfrac{v}{c^2}\right)u_x}$$

$$U_y^{'} = \dfrac{u_y}{\gamma\left[1 - \left(\dfrac{v}{c^2}\right)u_x\right]}$$

$$U_z^{'} = \dfrac{u_z}{\gamma\left[1 - \left(\dfrac{v}{c^2}\right)u_x\right]}$$

$$t^{'} = \dfrac{t}{\left[1 - \left(\dfrac{v}{c^2}\right)\right]}$$

to get the accelerations for mutually attracting frames we should note that the final event will be like that of a frame moving away from the original inertial frame. The original frame can be considered as stationary so that $U_x = $ **0 m/s**, $U_y = $ **0 m/s** , $U_z = $ **0 m/s** and **t = t'**. The final velocity therefore is Vn which is net of the two velocities and $U_x^{'}$ under Lorentz Velocity Transformations equations. The acceleration will therefore be $a_x = v/t$ which is also the net acceleration of the two accelerations. Under Lorentz Velocity Transformations; $\mathbf{a_x} = U_x' / \mathbf{t'}$ which reduces to;

$$a_{xn}' = \frac{\dfrac{u - V_n}{1 - \left(\dfrac{v}{c^2}\right)u_x}}{\left[1 - \left(\dfrac{V_n}{c^2}\right)\right]} \Bigg/ \dfrac{t}{}$$

$$a_{xn}' = \frac{-V_n\left[1 - \left(\dfrac{V_n}{c^2}\right)\right]}{t}$$

Where a_{xn}' is the net acceleration along x axis. The negative sign indicates repulsion since the equations are derived from a receding inertial frame. The negative also shows slowing down in our case. When the speed is slow the equations reduces to usual equation of acceleration. When the speed is equal to the speed of light, the acceleration is zero which consistence with the fact that speed of light is constant.

$$a_{yn}' = \frac{\dfrac{u_y}{\gamma\left[1 - \left(\dfrac{V_n}{c^2}\right)u_x\right]}}{\left[1 - \left(\dfrac{V_n}{c^2}\right)\right]} \Bigg/ \dfrac{t}{}$$

$$a'_{yn} = \frac{u_y}{\left[1 - \left(\dfrac{v}{c^2}\right)\right]\gamma}$$

,

$$a'_{zn} = \frac{u_z}{\gamma\left[1 - \left(\dfrac{V_n}{c^2}\right)u_x\right]} \bigg/ \frac{t}{\left[1 - \left(\dfrac{V_n}{c^2}\right)\right]}$$

$$a'_{zn} = \frac{u_z t}{\left[1 - \left(\dfrac{V_n}{c^2}\right)\right]\gamma}$$

For attracting inertial frames the acceleration will be;

$$a'_{xn} = \frac{V_n\left[1 - \left(\dfrac{V_n}{c^2}\right)\right]}{t}$$

$$a'_x = a_n\left[1 - \left(\dfrac{V_n}{c^2}\right)\right]$$

Relativistic Force

For objects attracting we can find the

relativistic net force by considering one frame stationary. So from $F_n = F1 + F2$ which is $F_n = m_1a_1 + m_2a_2$ then fixing one frame gives as its acceleration as zero ie $F_n = m_1(0) + m_2a_2$ which reduces to $F_n = m_2a_2$. But a_2 will now be the net acceleration so that we get $F_n = m_2a_n$.

this net acceleration is $a^{nx'}$ under Lorentz transformation. Therefore net relativistic force is $F_{nx}' = m_1 a^{nx'} = \dfrac{mV_n\left[1 - \left(\dfrac{V_{nx}}{c^2}\right)\right]}{t}$

$$F_{nx}' = m_1 a_{nx}\left[1 - \left(\frac{V_{nx}}{c^2}\right)\right]$$

CHAPTER 4

Galileo was Wrong: Masses do not Fall at the Same Rate

Net Acceleration due to Gravity, g_n

From our section on attractive bodies, if we

let $a_o = g_e$ and call it the acceleration due to gravity of the earth, and $a_e = g_o$ and call it the acceleration due to gravity of the object then the **net g** is;

$$g_n = g_e + g_o$$

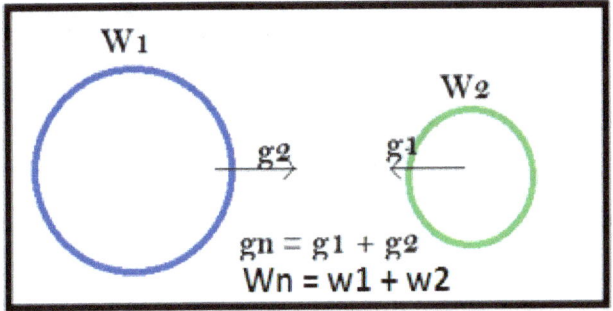

So for an object M1 coming from outside the other object's field, it will accelerate towards M2 with acceleration of a1, and M2 will accelerate towards M1 with acceleration of a2 and the net g is the sum of the two gs.

The net g experienced by an object when falling on earth or any other planet is also the net g experienced by the earth or planet. We must no longer talk about a specific g of the earth or object. The only g we can talk about is the g net of each earth/planet-object system. The g nets are specific to the masses and not general.

Objects do not Fall at the Same Rate

Since net g is not constant and is the sum of g of the earth/planet and the g of the object, massive objects with more g_o will move closer to the earth/planet faster than a less massive object with less g_o.

So objects do not fall to the ground at the same rate for an observer on that ground. But because we deal with mostly tiny masses, ordinary objects, compared to the earth or planet, the difference in their masses are negligible and hence their net gs differ slightly as to be imperceptible to the naked eye. This net g is about 9.81N/kg at sea-level. So the g of the earth is stated as always $9.81m/s^2$.

Galileo Galilei was wrong. His famous experiment was an oversimplification which led to error because it was not accurate enough. In the absence of air drag, heavier masses/objects will fall faster than less massive objects.

Net Force due to Gravity, F_n

Newton said that all objects attract each other i.e. all objects have gravity. When placed near each other, the objects will therefore fall into each other. If object 1 has force 1 and object 2 has gravity force 2 then the net force will be = F1+F2. The net force therefore between the two is greater than their individual gravitational forces.

So $\boxed{F_n = F1 + F2}$

This Fnet will be different for two objects of different masses brought near the same object because of differences in their forces of gravity.

The Complete Definition of Weight

The force the earth exerts on the object near the surface is called the weight. So F1 = W1 and the F2 can be called the weight of the earth on 'planet object' so that F2 = W2. So

$$\boxed{F_n = W1 + W2}$$

But w =mg so we get;

$$F_n = m1g2 + m2g1$$

or $\boxed{F_n = m_e g_o + m_o g_e}$

tor objects near each other.

But the force between two objects is also given by $F_g = G\dfrac{M_1 M_2}{r^2}$

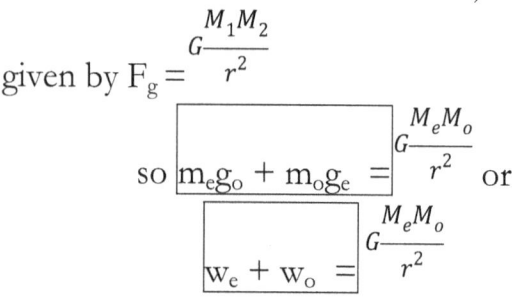

so $\boxed{m_e g_o + m_o g_e} = G\dfrac{M_e M_o}{r^2}$ or

$\boxed{w_e + w_o} = G\dfrac{M_e M_o}{r^2}$

The definition of weight is given as; $M_o g_e$ and is equated to $G\dfrac{M_e M_o}{r^2}$. But this definition ignores force of the object on the earth. The complete definition of weight, therefore, should include the force of the object on the earth or planet. The complete definition should be the sum of the two forces involved in the 'tug-of-war'. It should be; $m_e g_o + m_o g_e = G\dfrac{M_e M_o}{r^2}$

i.e. $\boxed{W = m_e g_o + m_o g_e}$

OR

$\boxed{W = W_e + W_o}$

When we measure weight using a spring balance we are actually measuring this net force or net weight. The specific weights of the object alone or earth alone can never be found i.e. w_e and w_o. When we measure the weight of the object we are simultaneously finding the weight of the earth on that object. It is like the faces of a coin or the poles of a magnet; they can never be separated to stand on their own. We must be talking about the net weight therefore and not just weight. All weights we measure are net weights.

So the same net weight of the object on the earth is also the net weight of the earth on that object i.e. $m_e g_o + m_o g_e = W$.

It is not the weight of the Earth only or weight of the object only but of both.

For instance, an object of mass 60kg then, the weight on earth is 60 x 10 = 600N and the weight of the earth on the object is also 600N. This 600N is the weight of the object plus the weight of the object.

The weight of the object alone is not 600N. It is less than 600N and the weight of the earth on the object is not 600N but less almost zero.

This implies that we can get an approximate of the g of the object since the mass of the earth is known using the formula $W = mg$. since the g of the object is very small, we can use this standard definition of weight. So $600 = m_e g_o$ and since the mass of earth has factor 10^{24}kg, we get g_o in factor of 10^{-24} N/kg.

Since it depends on the net g, it is possible for the complete weight of the earth or any other astronomical body to be small or even zero. Since the pull of the object on the earth is small, the earth accelerates slightly or remains stationary. The force/weight, therefore, since it depends on the acceleration, will also be small or zero in spite of the large mass of the earth.

Conservation of Position

For objects in contact with each other and in the same field, when the object is lifted up from the earth, the earth also moves backward due to the reaction force by Newton's third law. When the objects falls back, the earth also moves towards the object due to the gravitational force. The two objects will meet again on the same position. So for an isolated system position of the objects is conserved.

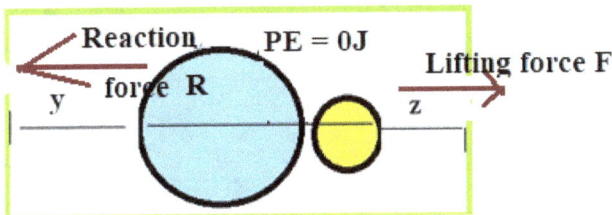

The PE of the two objects in relation to each other is zero when the objects are in contact. When the smaller object is lifted up, it gains potential energy PE1 in relation to the bigger object and distance Z.

The bigger object gets PE 2 in relation to smaller object and distance Y and $PE2 = M_e g_o y$, $PE1 = M_o g_e z$, this can be called positional PE. It is different from the PE after the final destination of the two objects.

When the objects fall back into each other, the PE is again back to zero. So the energy of the system is conserved and the position of the system is conserved i.e. the earth object system moves back to the original position.

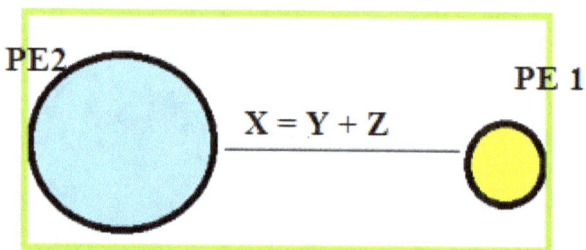

The Flaws of Galileo's Experiment

Galileo's experiment on Tower of Pisa had four flaws;

i) the earth-object system was to be isolated for each object so that the gravity of one does not interfere with the gravity of the other. The gravities of Galileo's feather and stone acted as that of one mass and force in relation to the earth. They therefore fell at the same rate. If they were separated and shielded from each other, Galileo could have observed that they fell at different rates

iii) the difference between the two masses was not sufficiently great to show the difference in the rate of fallings. Because Galileo dropped very tiny masses with very small g in comparison to the earth, the net g for each dropping differed very slightly as to be imperceptible. If he had done the same experiment with a very tiny object like an atom and another massive object like the moon, the moon would have fallen faster than the atom i.e. g_n moon and earth system will be greater than g_n for an atom and earth system. The moon has g 1/6 th that of the earth and an atom must have g significantly smaller than that of the moon. So g net of moon earth system will be greater than that of atom-earth system. In short, the moon will fall faster to the earth than the atom whose g net is less. The moon, if brought near the earth, will fall to earth at about 9.81m/s^2 and the earth is falling into the moon at about 1.64 m/s^2. We will observe this as a net g greater than 9.81 m/s^2 i.e. 11.5 m/s^2 because we are part of the earth. Because they will meet faster, in a shorter time, when we plug in that time into velocity equation we will get a g bigger than normal.

From the equations of motion, the time of falling for objects into the earth will be;

$$t_o = = \frac{V_e\sqrt{g_e}}{g_e\sqrt{g_o}}$$

t_o, v_o being time for object and velocity of object. The time for earth falling into object will be;

$$t_e = = \frac{V_o\sqrt{g_o}}{g_o\sqrt{g_e}}$$

We do not have the sense of us moving into the opposite direction. To us only the moon or falling object will seem to be moving/falling.

For an observer on the earth or planet, objects therefore do not fall at the same rate. It is like jumping up to catch a falling ball or object; the object is moving towards you at about 9.81 m/s² and you are also accelerating towards the object. You therefore meet the object earlier than if one did not jump. The net g of the jumping person and the falling object is therefore greater than 9.81m/s² for an observer in space who does not see the ground as a reference point of the two motions. To that observer objects fall at different rates. We have been wrong for more than 400 years.

ii) The distance from which the objects were dropped was also not sufficiently great to show the difference in the rate of fallings. If two smaller objects smaller than the earth by a factor of a billion and hence their gravities were of a factor of let's say 10^{-6} but differed slightly (like Galileo's feather and stone or Scot's feather and hammer), then by; $y = -0.5gt^2$, they would fall a distance of ;

6 x 10^{-5}m in 120 sec i.e. 2 minutes, and

6 x 10^{-4}m in 1200 sec or 20 minutes, and

6 x 10^{-3}m i.e. 0.006m or 0.6cm or 6mm in 200 minutes or 3.3 hours

So Galileo would have required a very high height or very big difference in the masses of the objects to detect the difference in the falling rates between them. A height which would give a difference of 0.6cm or 0.006m would require a height of;

$y = 0.5(9.81)(12000)^2 = 706,320,000$m

or 706,320km.

So just to observe a difference of 6mm in the falling rates for small objects would have required a height or ramp more than 700 thousand km up into space, and a waiting time of more than 3 hours to see just one falling object reach the ground.

In short, there was a slight difference in the rate of falling of the feather and stone but Galileo could not easily observe it. With the speed at which the object hit the ground, Galileo could not have possibly differentiated the rates with his naked eyes. But to his credit in his rigorous experiment, he managed to slow the speed using ramps. But they were not sufficiently high and the objects did not sufficiently differ in mass. They worked correctly in determining that the rate is exponential in form but it is not the same for all objects.

iv)Galileo did not know Newton's law of Universal Gravitation. He did not know that the earth was also falling into the feather and the stone. He did not take in consideration the movement of the earth into the opposite direction.

Newton, too, could have proved that Galileo was wrong. But I guess he couldn't fathom challenging the great master. Newton could not see the implications of his own law of Universal Gravitational in relation to rate of falling bodies.

Factors that Determine Size of g_0

It seems that the accelerations due to gravity of objects and planets are directly proportional to the masses and density of the objects or planets. For instance g of earth is 9.81m/s^2, for the moon it is 1.64m/s^2.

$$\text{So} \quad g_o = k\rho m_o \,, \quad g_o = k\frac{m_o}{V}m_o$$

$$, g_o = km_o^2/v, \text{ but } V = \frac{4}{3}\pi r^3,$$

$$\text{so } \boxed{g_o = k3m_o^2/4\pi r^3}$$

k is constant of proportionality which can be called constant of g or constant of weight. M_o should not be confused with mass of the object for which the weight is being found in relation to that planet.

Weight can also be found on any object and not only on a planet. So the notation of g_o and m_o is more appropriate where O stands for object. All this implies that weight must be actually lower than what we measure it.

This formula shows that the magnitude of g in the object varies directly as the square of the mass and inversely as the cube of the radius from the center of the earth i.e.

$$g_o = k_w m_o^2 / r^3$$

k_w being a new constant of the weight.

Estimating Gravities of Smaller Objects

The mass of the earth is in the factor of 10^{24}kg and that of the moon has factor of 10^{22}kg. If g is proportional to mass and density, and the gravity of earth is about 10N/kg then the g of the moon would be about 0.1N/kg. Taking the difference into view, we could multiply that by 10 to get about 1N/kg, which near to the real g of the moon. So the factors of the gs of the moon and earth in 10^0 and 10^1. We can also estimate the factor of gravity of smaller objects like a hammer or stone or 20 tonnes truck. If the feather has mass 5g or 0.005kg or 5×10^{-3}kg then its factor is 10^{-26} or 10^{-25} if we multiply by 10. If the hammer has mass 5kg or 5×10^0 kg (if it's the heaviest hammer). Then its factor is factor is 10^{-24} or 10^{-23}. We can see from this that the difference in the masses and hence in the gravities of the hammer and feather is miniscule. Galileo's experiment, therefore, was inadequate. For the 20 tonnes truck i.e. 20, 000kg or 2.0 x 10^4kg the factor is 10^{-19}. We can go on like this even with a million tonnes or 10^{10}kg ship. Its factor would be 10^{-14}N/kg. This shows that the objects we would consider heavy have very little gravity.

Contradictions if Galileo was Right

If we replace the g of Jupiter or the sun or indeed the whole universe, whose gs must be proportionally larger than that of the moon , where the g_m is in the formula $9.81 = g_m + g_e$ then we get a real g which is negative or opposite of what we would expect. This means the opposite effect of a falling object i.e. slowing down instead of speeding up as an object falls towards the ground. Since by definition the gravitational force is attractive, we would not expect it to do otherwise.

The net g must therefore be always positive for our formula to stand. It must be changing for us to get a positive real g of any planet or star near the surface of the earth. That is; objects must be falling at different rates. So we can longer talk about a constant g of the earth or planet. The only g we can talk about is the g net and specific gs of each earth/planet-object system.

Nevertheless, we can say that for astronomically small objects i.e. smaller than the moon and larger than an atom, the difference in their net gs is so small that we can say they always fall approximately at the same rate.

Measurement of g

The measurement of g by pendulum is affected by this fact. The mass of the bob and it's gravity is tiny to show the real g of the system. More-over everything will be affected by the earth-universe system so the net g is the result of all the gravity of the universe and the earth.

Measurement of g and New Formula

The measurement of g by pendulum is affected by this fact. The mass of the bob and it's gravity is tiny to show the real g of the system. More-over everything will be affected by the earth-universe system so the net g is the result of all the gravity of the universe and the earth.

The period of the pendulum will be affected by the mass. It is no longer independent of the mass. If we replace $g_o = k\rho m_o$, into

$T = 2\pi\sqrt{\dfrac{l}{g}}$ we get $T = 2\pi\sqrt{\dfrac{l}{k\rho m}}$ but $\rho = \dfrac{M}{V}$, So

$g = km^2/V$

$T = 2\pi\sqrt{\dfrac{Vl}{KM^2}}$ but for a spherical bob $V = 4/3$

πr^3, $T = 2\pi\sqrt{\dfrac{4\pi r^3 l}{3KM^2}}$, $T = 2\pi(2\pi^2)\sqrt{\dfrac{r^3 l}{3KM^2}}$, $T = 4$

$\pi^3\sqrt{\dfrac{r^3 l}{3KM^2}}$, $T = \dfrac{4\pi^2}{9k^2}\sqrt{\dfrac{r^3 l}{M^2}}$ $T = \dfrac{4\pi^2\sqrt{lr^3}}{9k^2\ M}$,

$T = \dfrac{4\pi^2\sqrt{lr^3}}{9k^2\ M}$, $\left(\dfrac{2\pi}{3K}\right)^2$ can be called the Pendulum Constant P.

$$\text{So } T = P^{\dfrac{\sqrt{lr^3}}{M}}$$

This formula shows that the period varies inversely as the mass. We have been teaching that mass is not a factor, but when the mass is huge or when using a very sensitive pendulum, it is a factor.

The formula also shows that the period varies

directly as the radius of the bob. This is just a consequence of the inverse square law of gravitational force.

Galileo's hunch, which he was actually testing in the experiment at the tower of Pisa that the rate of falling is directly proportional to the density of the object was actually correct but the rate also depends on the mass. For the same substance with the same density the more massive object will fall faster than the less massive one. This is because the more massive object has more electrons and protons which cause more gravity force than for the less massive object.

The g of Two Objects Merged Together

If the mass of two objects M1 and M2 with gs, g1 and g2 merge together give M. Then M1 + M2 = M

But their new g, g_m, will not be the sum of their gs i.e. g1 + g2 $\neq$ g_m. This is because the new weight of M, WM, is not the sum of their weight i.e. W1 + W2 $\neq$ WM or M1g2 + M2g1 $\neq$ (M1 + M2)(g1 + g2) It does not mathematically stand. Instead, their gs will be;

$$(M1 + M2)g_m = M1g2 + M2g1$$

$$g_m = \frac{M1g2 + M2g1}{M1 + M2}$$

so the new g changes and must be smaller than the sum of their old gs.

A Suggested Experiment on the Moon for Nasa

Since there is almost a vacuum on the moon, little gravity, enough height and empty clear space, it is the best place to make a proper physics experiment on rates of falling objects. Projectiles can be fired one at a time from the moon to the some height and their rates of falling measured. At least five trials can be carried out and the average rates found.

Nasa should go back and perform that experiment again. David Scott's experiment was a dismal attempt. It cannot even be called a replica of Galileo's experiment; the height was just too short.

This time, 'David Scott' should choose the 'hammer' and 'feather' of much difference in mass. The objects should be propelled into the air one at a time to at least 5 different heights starting with at least 30m. This will prove that masses do not fall at the same rate. Nasa should conduct a real physics experiment but of epic proportions.

If a feather of mass 5g and g of 10^{-23}, and hammer of 5 kg with g of 10^{-25} are used. Then the feather and hammer will take 11.08 seconds to reach the ground of the moon if they fall at the same rate. If their gs are used in the equation, then the hammer will take; a time of factor 10^{-12} and the hammer

10^{-18} seconds. The difference will be in the factor of 10^{-12} seconds, since it will be between the -12th position and the -18th position. This time factor can be reduced by increasing the distance by a factor 10 and the mass with a g much bigger eg a that of a 20 tonnes truck.

On earth, Galileo's rigorous experiment of the ramps can be tried again with higher and longer ramps, maybe 50m, 100m or even 1km up in the air.

NASA should also carry a pendulum to the

moon with two different masses of a great difference in their masses. This will prove that the mass of a bob is a factor.

Pendulum in a Vacuum

An experiment can also be done on earth using pendulums in a vacuum. Since the slowing down of a pendulum is due to air drag, a pendulum in a vacuum will swing for ever or very long time until stopped by the small friction force between the particles in the string as it bends due to the swings.

Two pendulums can be constructed of the same length but different masses. The pendulums can be dropped from the same height. Since the bigger mass should drop faster, it will eventually overtake the smaller mass when it reaches the bottom. This overtaking may not be perceptible initially, but overtime the bigger g or acceleration of the bigger mass will make exponential increments in the difference of the two pendulums at the bottom. This will prove that the g is not constant and that the period is affected by the mass.

Anti-gravity, anti-weight and anti-g

From our section in chapter 3 on repulsion of bodies, if we have two repelling astronomical bodies then we can talk about gravity. The force of repulsion could be called anti-gravity, the net acceleration between them could be called anti-g and the force between them could be called anti-weight.

The Big-bang and Dark Energy

So if the big-bang really occurred then we have the force of antigravity pushing the galaxies away. The galaxies initially accelerated then reached a constant speed so that now those galaxies which are collinear to the point of the big-bang have constant difference in their distance between them. The galaxies which are not collinear have an angle between them. As the galaxies expand away from the point of the big-bang, the arc length between them increases though the angle between them remains the same. This increase in arc length makes the galaxies expand away from each other.

This expansion makes the universe seem as if it has energy, the dark energy. But this is the original energy from the bigbang.

The point at which the bang happened must be devoid of any matter and energy. It is therefore some sort of a white-hole we can call the anti-black hole. This anit-blackhole grows as the energy and matter continues to recede from it.

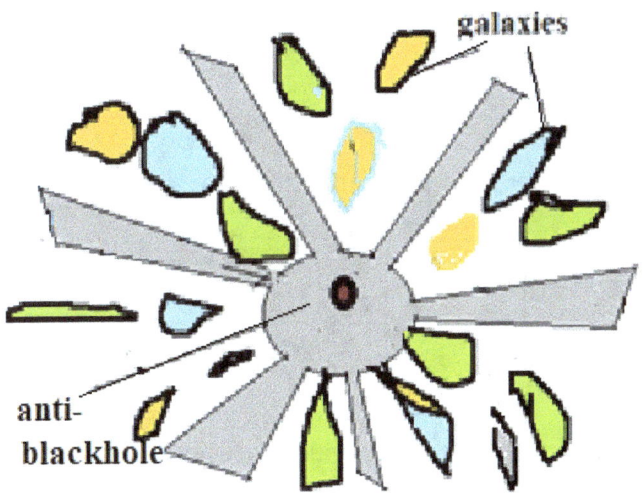

galaxies

anti-
blackhole

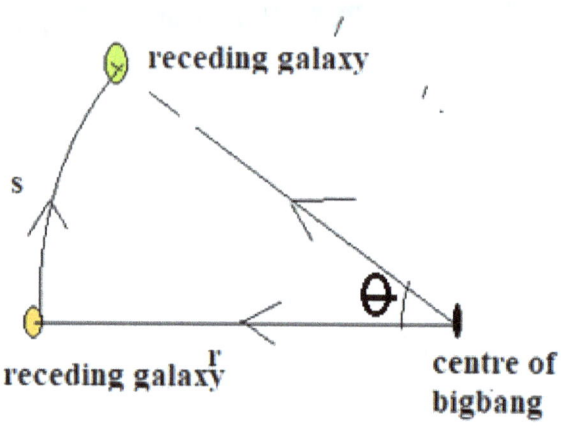

The Rate of Expansion

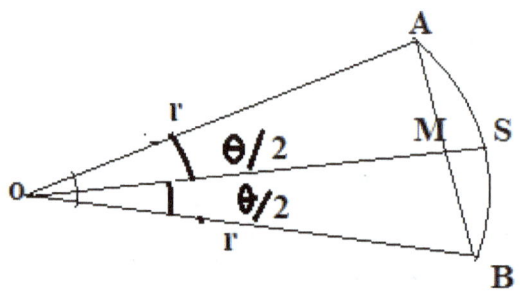

The distance of the galaxies from the center
of the bigbang assuming the galaxies were
travelling at the same speed is *r*. the angle
between them is θ. From definition of sine
then **Sin** $(\theta/2)$ = **AM/r**. Since the two

galaxies A and B are moving at the same speed, this ratio will be constant. The arc length, however, will be increasing and the distance AB will be increasing. This is the rate of expansion (R) between the two galaxies that are not collinear which Hubble observed. So from **AM = r sin ($\theta/2$)** we get, **AB = 2AM = 2r sin ($\theta/2$)** , **$d(2AM)/dt = 2dr/dt$ [sin ($\theta/2$)]** which gives net velocity between the two bodies/galaxies as ; $V_b = 2V$[sin ($\theta/2$)] which gives;

$$V_b = \pm 2v_g \sqrt{\frac{1 - \cos \theta}{2}}$$

where v_g is velocity from the center of the big bang.

The acceleration or rate of expansions (R) between the two galaxies is therefore; $V_b/dt = \pm 2v_g/dt \sqrt{\dfrac{1 - \cos \theta}{2}}$ which gives;

$$R = \pm 2a_g \sqrt{\frac{1 - \cos \theta}{2}}$$

This is also the formula for net acceleration between two bodies moving equidistant from a common point.

The Relativity of Apparently Mutually Attracting/Repulsing Objects

The above section gives an idea of how to treat apparent attraction or repulsion of two objects moving equidistant and at the same velocity from a common point.

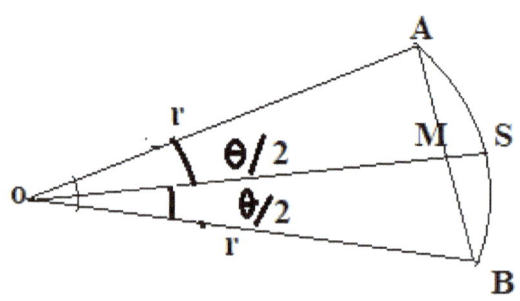

The distance of the two objects from the common point, assuming they are travelling at the same speed is **r**. The angle between them is θ. From definition of sine then; $\sin(\theta/2) =$ am/r, the apparent distance seen by an observer on one object is ;

$$r = d/\sin(\theta/2)$$

Where $d = AB = 2AM$. The observer on one object will see the other object as moving away from him/her or moving towards him at the velocity of ;

$$V_b \boxed{=} \pm 2v_g\sqrt{\frac{1 - \cos\theta}{2}}$$

where v_n is the apparent net velocity between the two bodies,
and acceleration of ;

$$a_n \boxed{=} \pm 2a_g\sqrt{\frac{1 - \cos\theta}{2}}$$

where a_n is the apparent net acceleration between the two bodies.

The Special-Relativity of Apparently Mutually Attracting/Repulsing Objects

If the object is moving from the point at near speeds of light then the formulae above will reduce to;

$$a_b = \pm 2a_g \sqrt{\frac{1 - \cos\theta}{2}}, \ \pm 2a_b' \sqrt{\frac{1 - \cos\theta}{2}},$$

$$a_b = 2a_n \left[1 - \left(\frac{V_n}{c^2}\right)\right] \sqrt{\frac{1 - \cos\theta}{2}}$$

for acceleration.

CHAPTER 5

ELECTROMAGNETIC WAVES AND ENERGY

Fields are Like Elastic Bands

Experience with repulsion of magnets shows that fields can be considered elastic and physical. When opposite fields are compressed together, they rebound. So fields are like stretched elastic bands connected at the center. When agitated a wave is set up in the field with energy of;

Planck's Formula; $E_n = nhf$

The tension of the field is like the tension of the elastic band and so is the magnetic force. Therefore each band has its own electrostatic elastic potential energy. In short, the electric potential energy is the electrostatic elastic potential energy

i.e. $EEP_f = \dfrac{1}{2} Kx$, were x is the distance of the field from the charge carrier and k is the constant of elasticity. So at each point

$$EEP = U, \ \frac{1}{2} Kx = QV$$

$$\boxed{V = \frac{Kx}{2Q}}.$$

k being the constant of elasticity or spring constant.

The electric field is the real ether.

The Electromagnetic Waves and Energy

The electric fields can be considered like elastic bands which have been stretched from the center of the electron or atom or object. Each position in a field will correspond to a single stretched elastic band a certain distance and direction from the centre. So each band or position has elastic potential energy.

The disturbance of the fields or bands, stretching them or compressing them, either increases the elastic potential energy or decreases it. This band disturbs the next band next to it, disturbing its elastic potential energy too. This band does the same to the next band. A wave therefore is created. This is the electromagnetic wave.

The field/band is stationary, it is an electric field. When it moves it becomes a magnetic field.

The Energy of the Fields

The energy as a result of the disturbance does not travel but oscillates within that band. When the field is stationary the energy is elastic electrical potential energy and the field is purely electrical field. When the field moves, this energy is converted to electric KE and the field is a magnetic field.

The disturbance in the energy reaches a certain point where it can be observed as light, heat, and gamma rays etc i.e. as EM waves. The EM waves do not travel to reach us; it is the disturbance travelling at the speed of light that reaches us just as the water particles do not travel in a water wave. What is disturbed in our eyes or ears when we sense light or heat are our electric fields in the eyes or skin.

Implications on KE Theory

To lose or gain energy is to have the fields disturbed around the particle. The particles are in constant motion because the fields are always losing/gaining energy/having their fields disturbed to a less or greater extent constantly.

Zeroth [Z] Waves and Eternal [E] Waves

The frequency of the disturbance of the bands produces each type of wave. Since the particles are always vibrating, they are always producing EM waves even when we can't detect those waves.

The waves from 0Hz to the radio waves which we can't detect can be called the Zeroth waves or infra- radio waves. All vibrating matter apart from producing sound waves also produces EM waves in form of infra-radio waves. It is therefore possible to detect sound without using a microphone but a receiver circuit tuned to an infra-radio wave frequency.

Since waves are produced when a particle is accelerating, a particle can accelerate indefinitely until it reaches the speed of light in one direction. For indefinite acceleration to occur the time must tend to zero and the change in velocity must also tend to zero.

 When the particle reaches the speed of light then it will no longer accelerate and the EM waves will cease being produced since the particle is in constant speed. The EM waves from Gamma to zero can be called Eternal waves, E waves or ultra gamma rays.

The electromagnetic spectrum is therefore

circular.

When the wavelength is zero and the amplitude infinitely long then we get the maximum frequency of EM waves.

When the wave length is very long and the frequency so low that the sine wave becomes a line then we get Z waves.

Energy, Frequency and Position

For the same amount of energy, the frequencies in each band will move from lower to higher, analogous to strumming strings of a guitar, since each band has its own length due to its distance from the centre. So the distance from the center correspond to the frequency and length multiplied by the frequency.

For the same amount of energy, Frequency is inversely proportional to distance of field from the centre i.e.

$F = k/x$, $k = Fx$, the k is the energy since this occurs for the same amount of energy.

Energy = frequency $\times$ radius of field

$$E = fr$$

When the radius is constant, that is; when energy is fluctuating within one field, the radius' magnitude equates to the magnitude of the Plank Constant h ie nhf = fr

$$r = nh$$

The Radius of the Photon

From $E = fr$, $r = E/f$. When the radius is constant, that is; when energy is fluctuating within one field, the radius' magnitude equates to the magnitude of the Plank Constant h i.e. nhf = fr

$$r = nh$$

This gives the radius of the specific electromagnetic field or photon. This radius corresponds to the energy as already stated. So the Gamma ray field has the greatest radius and the radio waves field has the least radius. If the n corresponds to the electromagnet spectrum, then n_6 for gamma radius will be n_6 x 6.62 x 10^{-34}j.s where 6 is the sixth position of the gamma rays in the EM spectrum, and

that of the radio waves will be
$6.62n_1 \times 10^{-34}$m

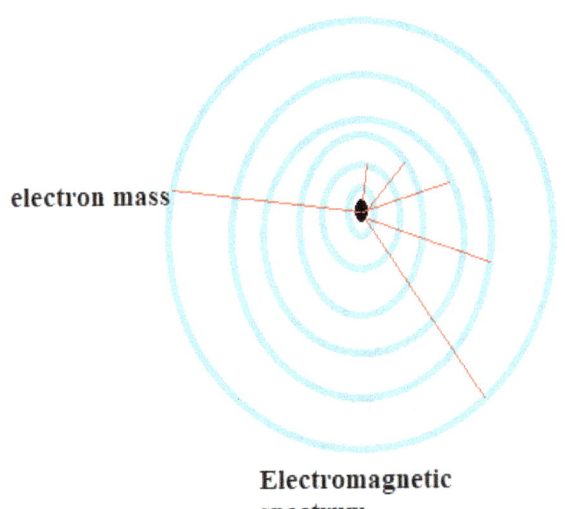

electron mass

Electromagnetic spectrum

The Fields Correspond to Each EM Wave

For the same amount of energy, the shortest field will produce the most frequency, and the longest field will produce the lowest frequencies. Since these frequencies correspond to types of EM waves, the levels of the fields correspond to EM waves. So we can have Z fields, radio fields, etc and so on up to gamma fields and E fields.

It is possible for one level of field to produce all these EM waves depending on the energy. So the longest field can produce the gamma rays if enough force disturbs it. This means these other fields will produce more than gamma rays i.e. they will produce Ultra-gamma waves or ultra—ultra gamma waves.

Redefining Matter

A piece of matter is like a solid core with a jelly mass around it. The core is the mass with its centre of mass and the jelly mass is the electric field.

Every object has the jelly mass/field around it even neutral objects. They have a thin layer of field due to polarization. This is equivalent to the Quantum Field though a field itself should not be described as a particle. It may behave like a particle but it is not a particle.

For a particle to interact with another particle,

it must have an electric field. For a particle to have directions and hence interact effectively, it must have the centre of mass. A particle with-out an electric field but with mass only is not matter, and it can never interact with other matter.

These fields contain energy which correspond to the EM waves. The agitation of these fields causes the energy to oscillate from potential to KE and in between and then back.

This oscillation or its frequency can reach our eyes or skin and the brain detects that frequency as light or heat. The energy is not ejected out. It just oscillates or transforms.

These fields can interact resulting in electromagnetism and gravity.

The electric fields spread out in space. On this electric field are levels of energy which correspond to the electromagnetic waves.

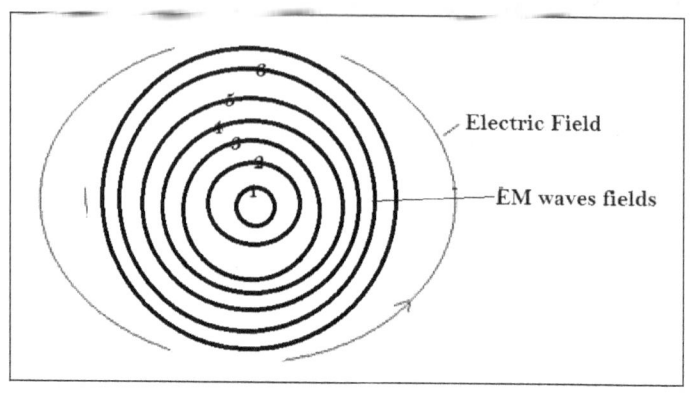

REFERENCES

1. Principles of Physics,Longman, London.

2. Serway R. and Jewett, J. 2004, 6th Ed, Physics for Scientists and Engineers,. Thomas Brooks/Cole,

3. Thornton, S. Rex, A. 2013 4th ed. Modern Physics For Scientists and Engineers, Brooks/Cole, Boston.

4. Zumdahl, S. 2003, Chemical Principles, 5th ed. Houghton Mifflin Company

5. Various Internet Resources